HOW TO READ

A WORKSHOP DRAWING.

By W. LONGLAND.

Being No. 6 of "The Home Worker's" Series of Practical Handbooks, edited by H. SNOWDEN WARD.

British Library Cataloguing-in-Publication Data
A catalogue record for this book is available from the
British Library

Technical Drawing and Drafting

Technical drawing, also known as 'drafting' or 'draughting', is the act and discipline of composing plans that visually communicate how something functions or is to be constructed.

It is essential for communicating ideas in industry, architecture and engineering. The need for precise communication in the preparation of a functional document distinguishes technical drawing from the expressive drawing of the visual arts. Whereas artistic drawings are subjectively interpreted, with multiply determined meanings, technical drawings generally have only one intended meaning. To make the drawings easier to understand, practitioners use familiar symbols, perspectives, units of measurement, notation systems, visual styles, and page layout. Together, such conventions constitute a visual language, and help to ensure that the drawing is unambiguous and relatively easy to understand.

There are many methods of constructing a technical drawing, and most simple among them is a sketch. A sketch is a quickly executed, freehand drawing that is not intended as a finished work. In general, sketching is a quick way to record an idea for later use, and architects sketches in particular (in a very similar manner to fine artists) serve as a way to try out different ideas and establish a composition before undertaking more finished work. Architects drawings can also be used to convince clients of the merits of a design, to enable a building constructer to use them, and as a record of completed work. In a similar manner to engineering (and all other technical drawings), there is a set of conventions (i.e particular views, measurements, scales, and cross-referencing) that are utilised.

As opposed to free-sketching, technical drawings usually utilise various manuals and instruments. The basic drafting procedure is to place a piece of paper (or other material) on a smooth surface with right-angle corners and straight sides – typically a drawing board. A sliding straightedge known as a 'T-square' is then placed on one of the sides, allowing it to be slid across the side of the table, and over the surface of the paper. Parallel lines can be drawn simply by moving the T-square and running a pencil along the edge, as well as holding devices such as set squares or triangles. Other tools can be used to draw curves and circles, and primary among these are the compasses, used for drawing simple arcs and circles. Drafting templates are also utilised in cases where the drafter has to create recurring objects in a drawing – a massive time-saving development.

This basic drafting system requires an accurate table and constant attention to the positioning of the tools. A common error is to allow the triangles to push the top of the T-square down slightly, thereby throwing off all the angles. Even tasks as simple as drawing two angled lines meeting at a point require a number of moves of the T-square and triangles, and in general drafting this can be a time consuming process. In addition to the mastery of the mechanics of drawing lines, arcs, circles (and text) onto a piece of paper – the drafting effort requires a thorough understanding of geometry, trigonometry and spatial comprehension. In all cases, it demands precision and accuracy, and attention to detail.

Conventionally, drawings were made in ink on paper or a similar material, and any copies required had to be laboriously made by hand. The twentieth century saw a shift to drawing on tracing paper, so that mechanical copies could be run off efficiently. This was a substantial development in the drafting process – only eclipsed in the twenty-first century with 'computer-aided-drawing' systems (CAD). Although classical draftsmen and women are still in high demand, the mechanics of the drafting task have largely been automated and accelerated through the use of such systems. The development of the computer had a major impact on the methods used to design and create technical drawings, making manual drawing almost obsolete, and opening up new possibilities of form using organic shapes and complex geometry.

Today, there are two types of computer-aided design systems used for the production of technical drawings; two dimensions ('2D') and three dimensions ('3D'). 2D CAD systems such as AutoCAD or MicroStation have largely replaced the paper drawing discipline. Lines, circles, arcs and curves are all created within the software. It is down to the technical drawing skill of the user to produce the drawing – though this method does allow for the making of numerous revisions, and modifications of original designs. 3D CAD systems such as Autodesk Inventor or SolidWorks first produce the geometry of the part, and the technical drawing comes from user defined views of the part. This means there is little scope for error once the parameters have been set.

Buildings, Aircraft, ships and cars are now all modelled, assembled and checked in 3D before technical drawings are released for manufacture.

Technical drawing is a skill that is essential for so many industries and endeavours, allowing complex ideas and designs to become reality. It is hoped that the current reader enjoys this book on the subject.

HOW TO READ
A WORKSHOP DRAWING

WHILE no one looking at the figure on the left in
Fig. 1 has any difficulty in recognising it as the
picture of a pulley, not everybody looking at the
other two figures would recognise that they represent the same thing. It is a fact, however, and
it is how a pulley would appear in a workshop
drawing. But why should two figures be drawn
when one, apparently, would not only suffice, but
besides would be more understandable? As a
matter of fact it is easier, and therefore takes less
time to draw the two figures to the right than the
one to the left; and what is much more important,
it is only by drawing the pulley as shown in the
two figures that its true shape and size can be
ascertained. Who could tell, by examining the
left-hand figure, that the pulley is circular, except
he knew it beforehand? It might be narrower
than it is high. True, circles when viewed sideway
appear oval, and so it might be guessed that the
figure represents something circular. Yet an oval

also appears oval, only more so, when seen sideway. Therefore it is only a guess when it is thought that a circular body is represented by the left-hand figure. There can be no mistake, however, when the figure furthest to the right is examined. This "elevation" represents the pulley as seen from the side. Apparently it has no width. The rim, arms, and nave can be distinctly seen, but there is nothing to show that the rim and the nave stand out from the arms. Plainly, then, this figure does not give all the information required to understand the shape of the pulley. The middle figure shows the pulley as viewed from the front; its width is given, but there is no means of discovering that it is round. Indeed, it may be thought that this does not represent the pulley as seen from the front, for there is no mistaking the fact that a real pulley is round when looked at from the front, even if neither of the sides can be seen. This is true when the light and shade gives this information, and not the lines which bound the surface. These are arranged as shown in the middle figure, and are all that can be seen from the front. Here again only a small part of the information required about this pulley is given, for only the width is added. But taken with the figure to the right much can be learned about the size and shape of this pulley. For the diameters of the rim both inside and outside, of the nave, and of the hole in the nave are there to be measured if required. The widths of the arms and of the rim can be obtained, but not the thickness of the arms nor the length of the nave. The means of obtaining these will be shown later. Drawings made like the two on the right supply more information concerning the true shape and the actual sizes of an object than the single figure on the left. On

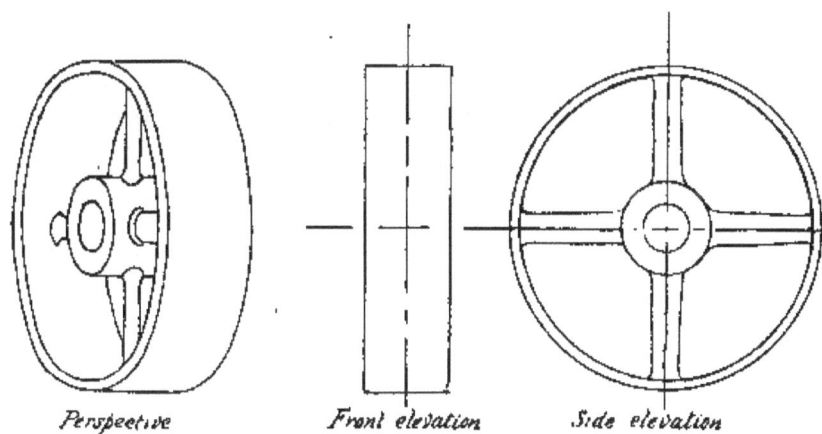

Perspective   *Front elevation*   *Side elevation*

Fig. 1.—A Pulley.

the other hand, however, that on the left gives a better idea of the general form of the object than the two on the right, even when taken together.

To ascertain the true shape and size of an object, at least two views of it must be shown. The meaning of "a view" as used throughout this book is a drawing made of the particular object in question when looked at from such a position that only two of its three dimensions can be seen. Thus its length and breadth may be seen, but not its thickness also, or its length and thickness may be seen, or its breadth and thickness only. At any rate, a drawing showing any two of these is called a view. There are three principal ways in which the body can be looked at so as to obtain a view, namely, from the front, from the side, and from above, looking downwards. The front view is called the "Front Elevation," or the "Elevation"; that seen from the side is called the "Side Elevation," or "End Elevation"; and that seen from above, looking downwards, is called the "Plan." If a view is required from below, looking upwards, then this is called a "Plan from Below"; but this is very rarely required. Fig. 2 represents an ordinary taper key. The elevation is drawn as seen from E in direction of arrow. It shows the length and thickness of the key, but not its breadth. The plan, drawn as from P, gives the length with the breadth and not the thickness. The side elevation, drawn as from S, gives the breadth and thickness but not the length. Thus each view gives the dimensions in two of the three directions in which every solid body extends. The side elevation requires a little explanation, for it is not apparent at first where the four horizontal lines come from. The lowest line, of course, represents the bottom of the key. The next the top of the key, at the end remote from the head. The third from the bottom one is the line where the head commences, so that the distance between the second and third lines is the sloping top of the key. And, lastly, the fourth line represents the top of the head. In this view there is no indication that the third and fourth lines from the bottom are some distance behind the other two ; that it is so can only be gathered by look-

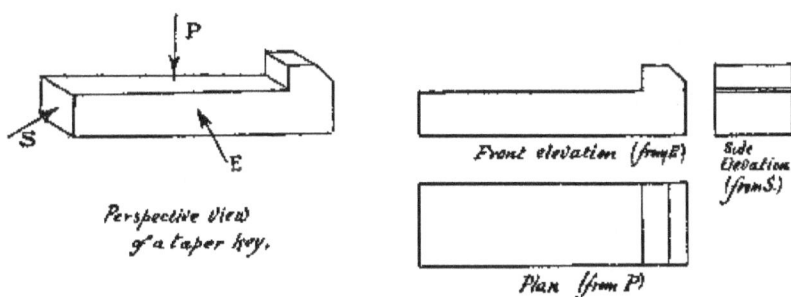

Perspective view
of a taper key.

Front elevation (from E)   Side Elevation. (from S.)

Plan (from P)

Fig. 2.

ing at the plan or the front elevation. This is hard to understand at first, but when one realises that only two dimensions can be obtained from any view, the practice of mentally combining two views is quickly learnt. Notice also that, taking all three views together, each dimension is repeated twice. Thus, if the plan and front elevation are alone drawn, the length, the thickness, and the breadth can be ascertained, and as -hat is all that is required, the side elevation is useless, and may be conveniently left out. Or suppose that the plan is not drawn : then from the front and side elevations the length, breadth, and thickness can all be obtained. Again, the plan and the side elevation taken together will supply these same dimensions, and the front elevation may be dispensed with. So, as a general rule, only two views are required, although, under some circumstances, it is necessary to have all three.

Of the two elevations, which is to be called "the front" and which "the side"? This is very difficult to say, for, taking the taper key in Fig. 2 as an example, if it is turned on its end, then, keeping the same view as the front elevation, the side elevation is that which is now shown as the plan. Or, turning it on its side, the front elevation and plan change places. Thus confusion may arise if there be no rule concerning them. This, then, may be taken as the rule—the front elevation is that view which has another view either below or above it (this other view being called "the plan"). And if there are two elevations but no plan, then that which has the greater length may be looked upon as the front elevation. Notice that in the rule just given the plan is said to be placed either below or above the front elevation. In England it is customary to place the plan below the front elevation—that is to say, the view of the top is put underneath the view of the front. Similarly, the view of the right-hand side of the object should be put to the left of the view of the front, and that of the left-hand side should be to the right of the front view. This is illustrated in Fig. 2, for here the view seen when looking from S is placed to the right. Although this is where it should be if the plan is below the front elevation, yet sometimes the side elevation is placed on the same side of the front elevation as that from which it is seen. If this were done in Fig. 2, the side elevation would have to be placed to the left of the front elevation. This position for the side elevation is helpful in reading the drawing. For, again referring to Fig. 2, it is seen that if the side elevation remains where it is, then the eye has to travel from the extreme left of the front elevation across this view and also the intervening space between the elevations to find corresponding points and lines on the two views. While, if the side elevation be transposed to the left of the front elevation, it is only across the space between the views that the eye has to traverse to find corresponding points. Although in this case the advantage is not great, yet, when the front elevation is large and covered with lines, the arrangement of the side view on the side from which it is seen is distinctly useful for rapid and accurate reading of a drawing. The same may be said about the plan.

For here the eye has to cross the elevation from the top to the bottom, and then the space between these views before points in the elevation can be connected with points in the plan. So, then, it would seem an advantage to place the plan above the elevation instead of below it, and at the same time place the side elevation on that side from which it is viewed. This is the method adopted largely in America and to a less extent in England. So, then, there can be these arrangements of the views :—

(1) The plan above the front elevation and the side elevation placed on the side from which it is seen.

(2) The plan below the front elevation, with the side elevation placed on the opposite side to that from which it is seen.

(3) The plan below the front elevation and the side elevation on that side from which it is seen.

Carefully notice this last method. It combines parts of the first two methods. Thus the plan is placed in the same position as it is in (2), while the side elevation has a corresponding position to that in (1). This method is widely used, and care must be taken not to confound it with (2).

It has been pointed out that two views are required to give all the dimensions of an object. Fig. 3 illustrates how essential it is to keep these views under observation at the same time. Here links of totally different forms are seen to have similar elevations. The first, (a), has a single boss on each end. The elevation of these bosses are seen as circles, while the rest of the link is shown in elevation by lines joining these circles, one on the top, the other on the bottom. In the plan the bosses are seen as rectangles, one at each side, standing out from the rest of the link. The

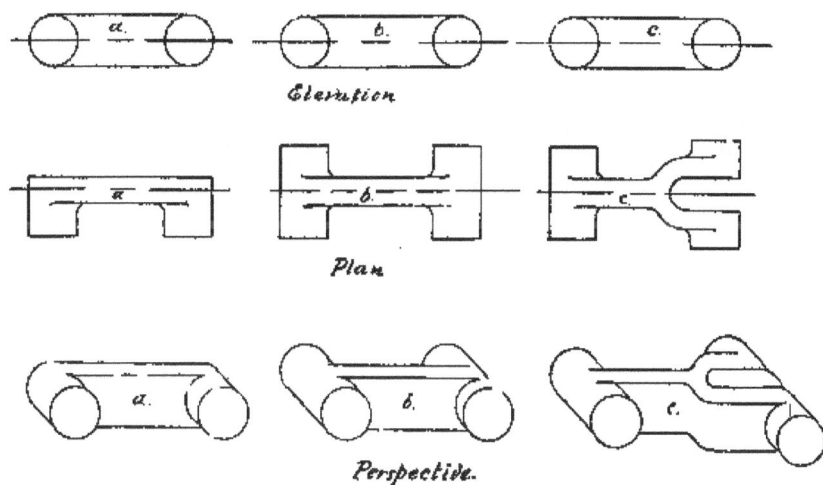

Fig. 3.

next, (b), although it has bosses on both sides at each end, yet in the elevation only those bosses which are on the one side can be seen. The plan, however, shows that there are bosses on the other side also, and thus distinguishes it from (a). In the case of (c), the plan shows distinctly that one end is forked and the other similar to that of (*b*), yet when it is viewed from the front, the elevation is exactly the same as that of both (*a*) and (*b*). In Fig. 4 the same principle is illustrated. Here are shown a pin and washer. One is a cylinder projecting out of another cylinder, while the other is a cylinder with a hole through it. And although their front elevations are entirely dissimilar, their side elevations have nothing to show that the inner circle

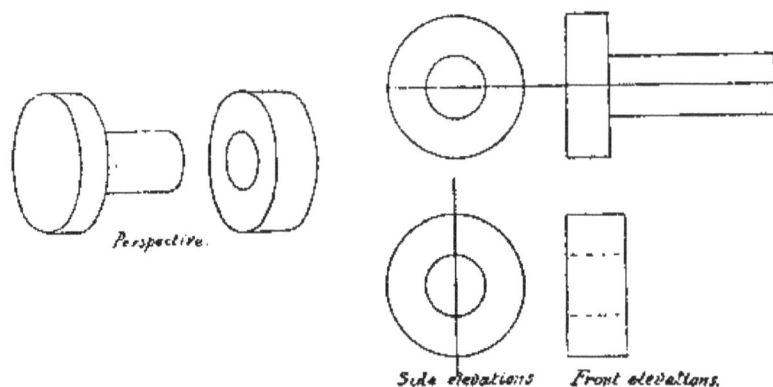

Perspective.

Side elevations   Front elevations.

Fig. 4.

represents a hole in one case and a cylinder in the other. It may be mentioned here that the plans of both are exactly the same shape as their respective elevations. Enough has now been given to show that one view must be used to interpret another.

In "Workshop Drawings" there are certain kinds of line and certain combinations of letters of the alphabet which, when used, it has been

.agreed that they signify certain things, and those only. These are called "conventions," and some common ones will now be given.

Ls.	Angle irons.	G.F.	Grinding finish.
Bbt.	Babbit metal.	F.	Machined or
Bs.	Brass.		turned.
(B).	Bright, nuts, &c.	M.S.	Mild steel.
Bz.	Bronze.	F.B.	Polished.
C.I.	Cast iron.	T.	Tapping-hole.
C.S.	Cast steel.	T's.	Tee-irons.
G. 2·75″.	Gauge dimen-	Wh.M.	White metal.
	sion.	W.I.	Wrought iron.

f is sometimes used for "machined." It is put on the line which has to be thus dealt with. Sometimes the depth of the metal to be removed is indicated by some such means as the following:

r. = 1-16th inch of metal to be removed.

$f. = \frac{1}{8}$,, ,, ,, ,,

$F.$ = 3-16th ,, ,, ,, ,

$ff. = \frac{1}{4}$,, ,, ,, ,,

Then for nuts and the heads of bolts: *Hex.* means hexagonal; *sq.* means square; *ro.* means round.

We next deal with the conventional line shown in Fig. 5.

(a) The full line is used for outlines such as can be seen, while for those parts that are hidden a dotted line as (b) is used. Frequently hidden portions are omitted because dotted lines do not always add to the clearness of the drawing.

(c) A chain line, consisting of a long and then a short dash is used for center lines. Sometimes, however, these are in red, and then they are full lines.

(d) A chain line, consisting of a long and then two short dashes, is used for dimension lines. It is sometimes used also to show the path of a moving part, such as a crank-pin or a lever. When used for dimension lines it has arrowheads at its extremities, as shown in (e), also the dimension in inches, feet, or millimetres, as the case may be, is placed either in a gap in the line or just above it. This, then, means that the space between the points of the arrowheads measures the amount given. A single dash on the right-hand side of a numeral figure and just above it means that feet are indicated. Thus 2′ means 2 feet. Double dashes in the same position signify inches. Thus 2″ means 2 inches, while 2′ 2″ means 2 feet 2 inches. And, lastly, m/m beside a numeral figure means that the figure represents so many millimetres. Thus 292 m/m means 292 millimetres. This is the metric or French system of measurement, and dimensions are usually given in millimetres only and not centimetres, etc. The dimension lines are sometimes placed as shown in (f) (Fig. 5), that is, on

the view to be dimensioned; and sometimes, as
in (*g*), where the dimension line is taken up clear
of the view, and lines, called witness lines, extend
from that part of the view which is to be dimensioned to the dimension line. Both mean the
same thing, namely, that the size of the object

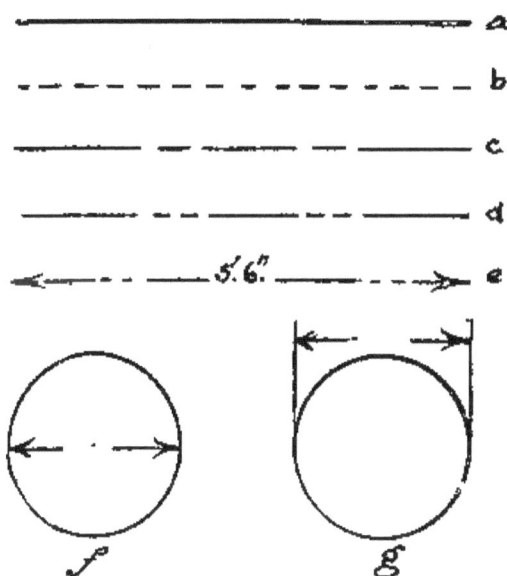

Fig. 5.—Line conventions.

shown is that given in or above the dimension
line. The dimensions are, perhaps, the most
important part of a drawing, and should be so
placed that the sizes of all parts can be readily
ascertained. But they must not be so crowded
together that one may be mistaken for another,
so sometimes it happens that dimensions which
might reasonably have been placed near together

are found some on one view and some on another. If, then, a dimension cannot be found where it is expected, do not conclude that it has been left out, but search for it on the remaining views. When it is necessary to see the shape of the inside of an object, it is usual to draw a view having a part cut away. Another convention is used to show which are the solid portions having been cut through. This consists of drawing a

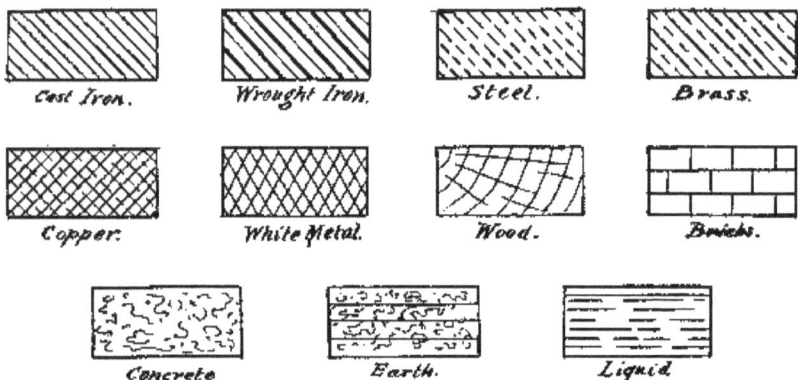

Fig. 6.—Conventional hatchings for materials.

series of sloping lines across these portions, and the name given to the lines is " hatching." In some cases advantage is taken of this to show the materials of which the parts are made by using a particular hatching for each material in common use. These are given in Fig. 6, and should be committed to memory. This is not difficult when a comparison is made of the hatchings used for materials which belong to family groups. Thus cast iron is represented by a series of parallel lines of uniform width, and wrought iron is distinguished by the lines being alternately thick and thin, while steel has a series of fine-dotted lines. Next, brass and copper are related to one another, and so their hatchings are something alike. Thus that of brass consists of alternate full and dotted lines, while that of copper is brass repeated, together with full lines sloping in the opposite direction. White metal is all alone. And

Fig. 6A.—Blocking in.

this is followed by the hatching for wood, which suggests that material. Brickwork also bears some resemblance to the actual thing. The hatching for concrete might be taken for a species of shorthand. Next comes earth, which is the hatching of concrete with horizontal lines across it, though why they should be thus shown somewhat alike cannot quite be understood. Lastly, the hatching for a liquid consists of horizontal lines of unequal lengths drawn across the vessel containing it. These special hatchings are not universally used, although hatching is always used when a cut portion of an object is to be represented, except when the drawing is coloured. When only one style of hatching is used for all materials, that shown for cast iron is nearly always adopted. If the portion which has been cut is very narrow, it is usually represented as shown at (a) and (b) in Fig. 6A. This is said to be "blocked in" and not hatched. Notice that a white line runs along the tops and the left-hand sides of all parts. The advantage of this is seen in (b), which is a section of a single-riveted lap joint. For if there were no white lines about each of the plates, which are riveted together, then the parts of the plates which overlap would show as one block of metal. Now, however, it can be distinctly seen that there are two plates riveted. The reason for putting the white lines just where they are is that the light is supposed to shine on the view at an angle from the left, and those parts on which it shines are distinguished by having the white line on them, while the other parts which are in the shade are without it. Thus the outside of the flange to the left in (a), Fig. 6A, has a white line along it, for it is fully in the light, but the inside of the same flange is without, because no light is upon it here. The top of the web is supposed to be in the full

light, for although the flange would in reality
screen a part of the web, yet it is assumed that it
does not, and so the web has a white line along
its whole length. The under side, however, has
the thickness of the metal between it and the
light, and so is in the shade, and, therefore, without a white line. Then the inside of the flange
to the right is in the light, and, therefore, is
bounded by a white line, while the outside, being
totally in the shade, has no such line about it.
Notice that here, again, the other parts of the
section are assumed to have no power of putting
any part of this flange in the shade, otherwise
the lower part of this flange would be screened
from the light by the web. This is also seen in
(b),:Fig. 6A. For here one plate covers up a part
of the other, and yet there is a white line between
them, which line belongs to the lower plate, and
has been shown as though the upper plate were
not there. The same can be said concerning the
white lines between the rivet and the plates.
The line to the left belongs to the rivet, but the
one to the right is on the plates, for the light is
supposed to shine on the right of the hole, while
its left is in the shade.

The supposition that the light shines on a view
from the left hand upper corner is sometimes used
in the case of ordinary plans and elevations.
With these a thin line is used to show where the

light is falling, while a thicker line shows the
parts in shade. It has this advantage, that one
can tell at a glance if the line in question is on the
outside or the inside. This is called shade lining,
and is illustrated in Fig. 7. Notice that here
again each part is independent of any other, so far
as light and shade are concerned. Thus the
shank of the pin does not throw any part of its
head into the shade. Compare Fig. 4 and Fig. 7
together, and compare the side elevations of the
former with the plans of (a) and (b). In Fig. 4
they are both the same, but in Fig. 7 the inner
circle is thicker on its lower part in (a) and on its
upper part in (b). The reason is that in (a) there
is a cylinder standing on another, and that the
light shines on the left-hand upper part of the top
cylinder and thus the right-hand lower part is in
the shade and is therefore thickened to show it;
while in (b) there is a hole through a cylinder and
here the light shines on the right-hand lower part
of the hole, leaving the left-hand upper part in the
shade, for which reason this is also thickened up.
The outer circles represent the outside of a
cylinder in both cases and are therefore both
thickened in the same part. Thus at a glance the
plan in (a), Fig. 7, can be seen to be that of a
cylinder standing on another, because the thickened
parts are both on the same side of the circles,
while the plan in (b), Fig. 7, is that of a hollow

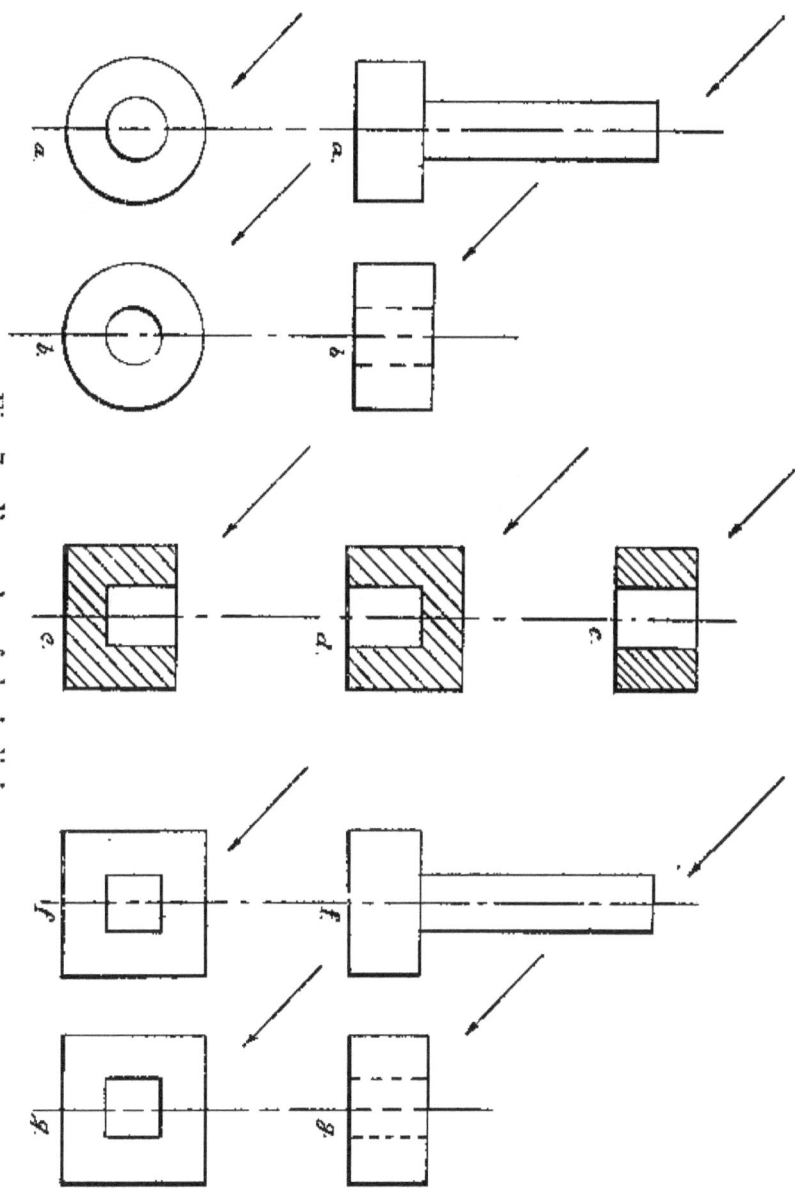

Fig. 7.—Examples of shade-lining.

cylinder because the thickened part of the inner
circle is opposite to that of the outer circle. At
(c), Fig. 7, is shown a view representing (b),
Fig. 7, cut in half, and the front portion
removed. It is called a section, but more
about this will come later. Notice however that
the cut parts are hatched; which is the conventional method of showing that it is cut, as has been
explained. Attention is to be drawn to the
disposition of the thick and thin lines. Thus
examining the vertical lines first, that furthest
to the left is thin, because the light is shining full
upon it; the next is thick being in the shade, by
reason of the thickness of metal between it and
the light ; the third from the left is thin being in
the light because it is the side of the hole remote
from the source of light ; while the fourth from
the left is thick being the outside of the cylinder
and thus having the thickness of metal between
it and the light. Notice that in the case of the
outside of an object the lines nearer the source of
light are thin, while those which are further away
are thick, and that in the case of a hole or
hollow the opposite holds good. Of the horizontal
lines (c), the top ones that are thin, and all those
which have solid metal between them and the
light are thick. It may be pointed out here that
(c), Fig. 7, would be the section of a cylinder
with a rectangular hole in it or a rectangular

mass with either a rectangular or a round hole in it. So that (c) being the section of all these different shapes there must be another view given before the actual shape can be ascertained. And this other view must be the plan, for either or both the elevations would be useless. Suppose now that the hole instead of going right through as in (c) only goes a part of the way as in (d) Fig, 7. There the same remarks apply to the vertical lines as were made about the same lines in (c). Of the horizontal lines, that across the top is thin because it is most certainly in the light ; next that across the top of the hole is thick because a mass of metal comes between it and the light ; and as the same applies to the line across the bottom of the cylinder it therefore is thick. If however, (d) be turned upside down, then, while the vertical lines remain as they were before, the horizontal ones have changed positions. This is shown at (e). Thus lines on any object may be thick or thin according to their position relative to the light. Going on to (f), here is an elevation similar to that in (a), but the plan shows that its shape is totally different. It can be seen that there are two rectangular prisms, one standing on the other (a prism is a solid, having as it ends equal, similar and parallel plane figures and those sides are parallelograms). It is said to be rectangular when the angles of the plane figures at its ends are all right angles. This is sometimes spoken of as a square, but this expression is only correct when all the sides of the end figures are equal. Returning to (*f*), it will be seen that the right-hand and lower sides are thickened, while the left-hand and upper sides are thin. The reason being that the rays of light are supposed to come from a source in the direction of the upper left hand corner as before.

In (*g*) the elevation is similar to that of (*b*), but there again the plan shows that the actual shape is very different. In this plan, because there is a hole through the mass of metal, top and left-hand sides of the inner rectangle are thick, while the other two are thin. The shade lines on the outer rectangle are in the same positions as those in the outer rectangle (*f*), and for the same reasons. Thus it is seen that shade lines often help considerably in the reading of a drawing. The reason why they are not universally used is that it is tedious work for the draughtsman to put them in all drawings, and takes an appreciably longer time.

From neither the plan nor the elevations can the shape of the inside be obtained, so that if this is required another view must be drawn. This is obtained by supposing the object to be cut in two parts and one of them removed; then on looking at the other the interior can be seen. This is called a section; if the cut is taken downwards the view is called a sectional elevation ; if taken across it is called a sectional plan. Not only does the section show the interior of an object, but sometimes it is essential to interpret the other views. In Fig. 8 are shown the front and side elevations of a steam engine piston, while beneath these are three sectional elevations, to each of which the front and side elevations might belong, for tracing out the corresponding parts in each section, and it will be found that the front elevations of them all are the same. Thus the hole through which the piston rod passes is seen as the inmost circle in the front elevation. Next the outside of the boss is another circle. Following this there is a circle representing the inside of the rim of the piston, while yet another circle represents the outside of the rim. All of these appear in the front elevation, but there is nothing to show where the web or spider is located. Neither does the side elevation give the required information, for the rim of the piston covers all else. Thus, a sectional elevation is required. The information might have been given by dotted lines on the side elevation, but it is much more clearly conveyed by the section. At first sight it is not quite clear why the horizontal lines joining the ends of the sections of the rim are shown on the lower views. Yet when it is remembered that these are views of a piston cut in half, and that there still remains a part of the piston behind the part which has been cut, it will be seen that these lines represent the back half of the rim, seen on the inside, however, and not on the outside, as shown in the side elevation. If these horizontal lines were left out of the views, then they would be sections and not sectional elevations. As a piston is round, these views may equally be called sectional plans, for the latter are exactly the same shape.

In Fig. 9 are shown the front and side elevations and also a sectional elevation of a shaft coupling. From the side and front elevations can be found all that is required concerning the exterior of the coupling. The diameter and the width of the flanges, the diameter and length of the bosses, the number and position of the bolts, all are to be found on these. But there are no means of ascertaining how the flanges are fixed to the shafts, nor indeed the length of the shaft in the flanges, so that, if these things are to be known, another view must be drawn. A section, either in plan or elevation, fulfils the requirements of this view. The sectional elevation shown (it might be a sectional plan also) has been so chosen that it passes through two bolts, and, besides that, it exposes to view the keys by which the shafts are fixed to the flanges.

Front elevation

Side elevation.

Three alternative sectional side elevations

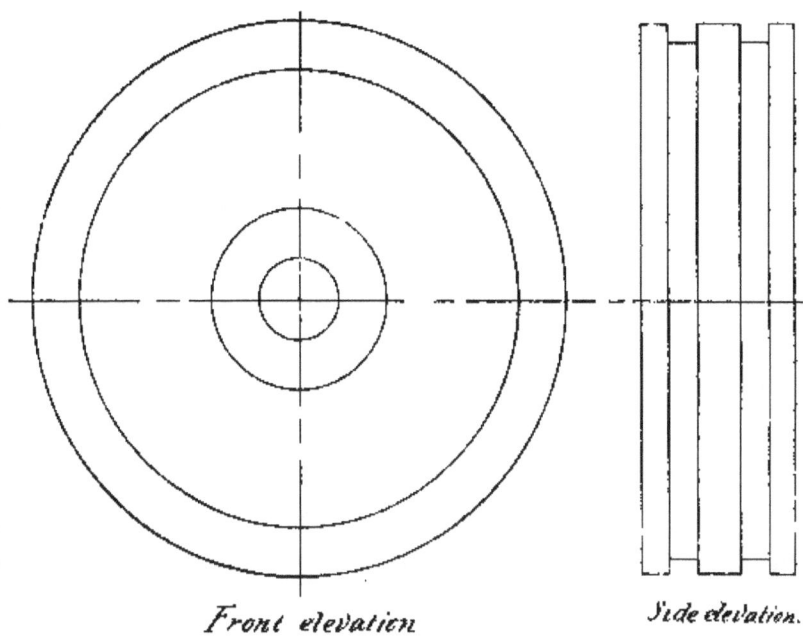

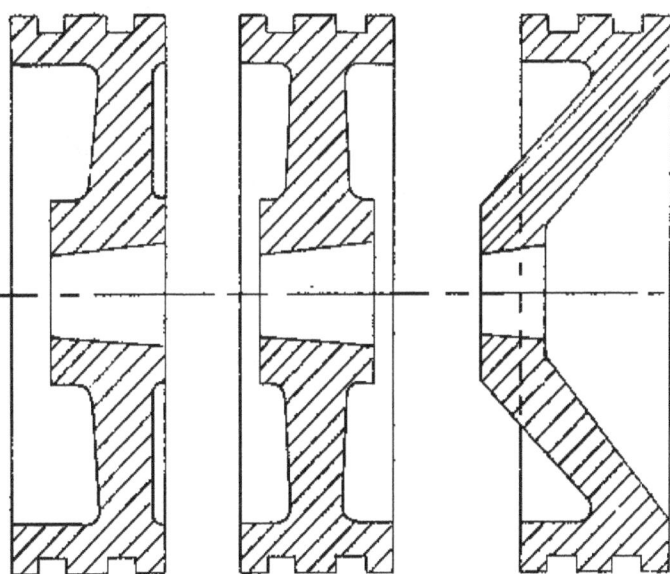

Fig. 8. Elevations and sections of pistons,

Although the section passes through the bolts, they are not shown in section, and the same can be said about the shafts, when it is remembered that all solids when cut are by convention to be hatched, which none of these are. Besides this, it will be seen on close examination that the bolt-heads and the nuts are shown exactly the same in both the section and the front elevation. Why is this? Why, if all these are in the section, should they not be shown in the conventional manner? Consider for a moment the value of a section. Is it not to show what is hidden by being inside? What is hidden inside, a bolt or a shaft? Truly nothing; therefore these are shown in outside elevation. So, then, bolts, nuts, shafts, spindles, rods, and all such things which can be understood even when not cut in half are left in elevation, while all the rest is in section. In this sectional elevation the keys can be seen, that on the left-hand shaft being on the top of it, and that on the right-hand one being at its middle. Also, one flange is recessed while the other has a projection fitting in this recess. This is to keep the shafts in line.

Occasionally, a part of an object is in section, as shown in the "detail" Fig. 9. This part has a wavy line round about it, which is to show that this is the only part in section. This is done when it is only a local detail, which it is desired to show.

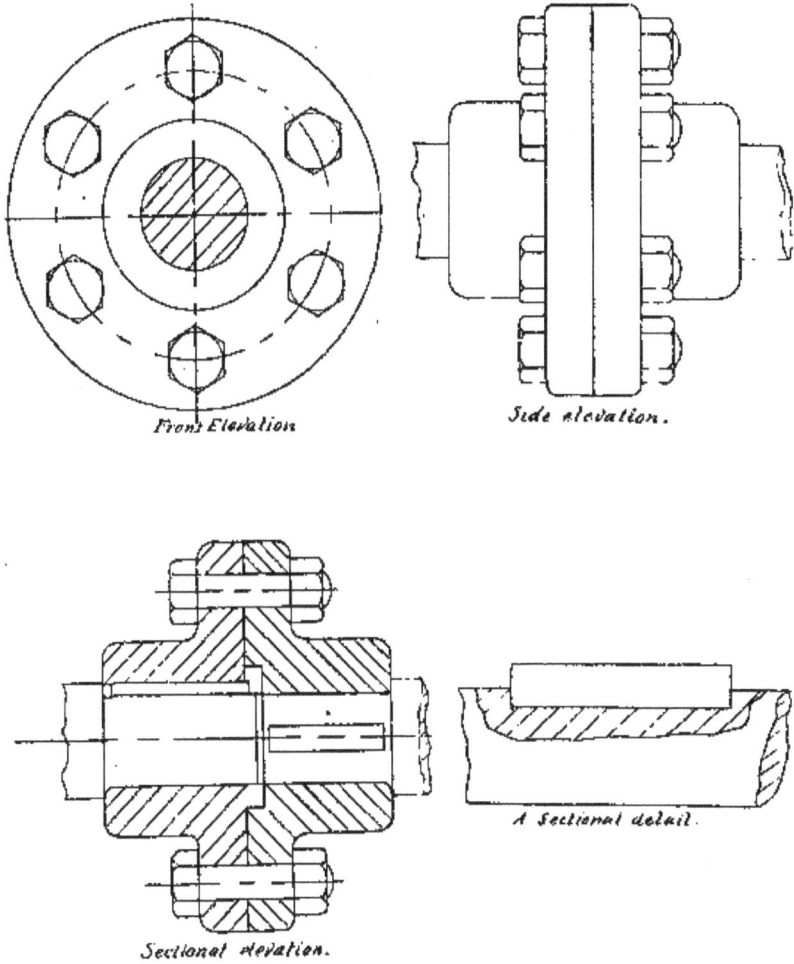

Front Elevation

Side elevation.

Sectional elevation.

A Sectional detail.

Fig. 9.   A shaft of coupling.

Thus, in this case the shaft is cut away round about the key, and thus shows how much it is sunk into the shaft; and as this is all that is needed, the rest of the shaft is left in elevation.

Notice that even here the key is not in section. Looking at the smaller detail it will be seen that one end of the shaft is an irregular oval, with hatching across it. This is often done to show that it is round. If it were square, there would be a narrow hatched rectangle on it. By doing this, it is unnecessary to draw a side elevation.

At (a), Fig 10, is another illustration of the use of partial sections. This is a front elevation of an eccentric sheave. It is divided into two parts along the line A A, so that it can be put on the crank shaft and these parts are then held together by means of the two pins and cotters, one pair on each side as shown. And here a partial section is all that is needed to show the shape and size of these pins and also how they are held in place.

Frequently by drawing a section of a part on its elevation a sectional plan or elevation is rendered needless. This is illustrated in (b), Fig 10, which shows the elevation of a wall bracket. On it is placed a section of the inclined part to show its shape. Another illustration of this is seen in (c), Fig. 10, which is the front elevation of a crane hook. The sections show very clearly its varying

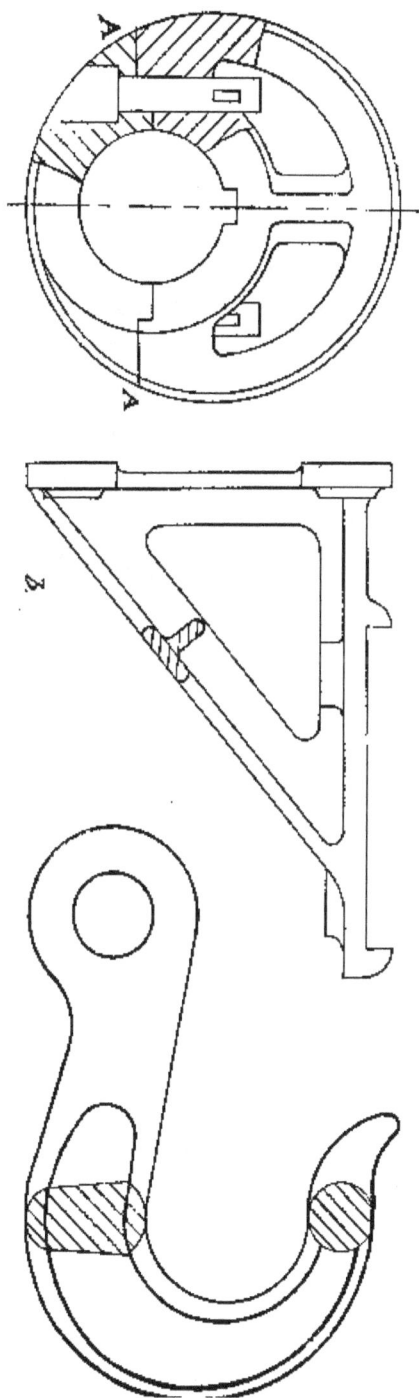

Fig. 10. Showing uses of partial sections.

form. Each is a section on the line dividing the section into halves. It must be clearly understood that these are sections, and not sectional elevations, and that they have no right to be where they are, but are placed on the elevation of the parts of which they are the sections for convenience.

On examining Fig. 11, which shows the elevations and sectional elevations of a friction clutch, it will be seen that both the side elevation and the sectional elevation are symmetrical about horizontal centre lines, that is to say in each view there are lines above the centre line which have lines below it corresponding to them in length and position, which simply means that twice as much work as need be has been spent upon the drawing; for if only half of each elevation were drawn combined as shown at (c), all the information given in the two views would be obtained. This means of economising space and labour is often adopted. A view such as this, means that if the section were extended below the centre line, the part above it would be repeated, while if the elevation shown below the centre line were extended above it, then it would also be repeated. When this is once grasped, there is not much chance of being led astray by one of these compound views. To the right of this compound view the front elevation of this clutch is shown. And it is noteworthy that the view is the front

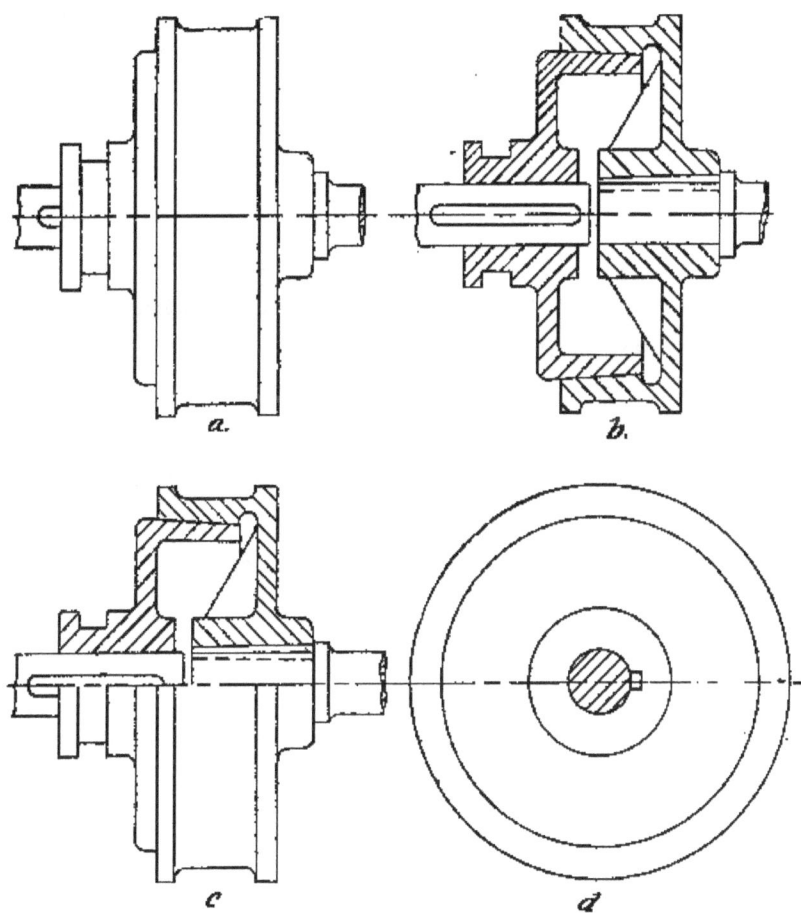

Fig. 11. A friction clutch.

elevation of the friction clutch as it actually is, and not a front elevation of the other view shown with it. For if it were, then the upper right-hand quadrant, (a quarter of a circle is thus named) should not be there. And this holds good for both plans and elevations given with sections; that the plan or the elevation is the whole plan or elevation of the body to be represented and not the plan or elevation of the section as drawn. This is perhaps at first rather confusing, yet it is quite reasonable. For the real shape of the object is required, irrespective of any sections which have been made of it.

All sections which have up to now been examined have been on a center line, that is to say the object has been assumed to be cut in half. If it were necessary, however, the section line could be taken anywhere else, in which case it is usual to write beneath it that the view is a section on such and such a line. If attention be paid to the position of the line on which the section is made no difficulty will be met with in understanding it. For all the remarks which have been made concerning a section apply to this one also. Occasionally, however, it happens that the line on which the section is made is not a straight one; which of course means that the sectional view is composed of two or more sections, one being further forward than the other, or, as it is said, in different planes. This is shown in Fig. 12. But before it is examined for the purpose of explaining the section, a few words may be given concerning what it is, and what it is used for. Firstly, then, it is a bearing for the crank-shaft of a large horizontal engine. And it must be understood that the steam acting on the piston, by means of the connecting-rod, forces the crank-shaft now to the right and now to the left, while all the time the crank-shaft is rotating in this bearing. This rotation of the crank-shaft causes the bearing to wear away, especially at those parts against which the shaft is forced. That the crank-shaft may run true, some means must be adopted to compensate for this wear. This is done by making the step in four parts, instead of two, which is the more usual way with bearings. The parts at the sides where the wear takes place are so arranged that they can be forced inward, and thus take up the wear on these parts. They are forced in by means of wedges by which they are backed. All this can be seen in Fig 12. It will be noticed that the steps are not divided into four equal parts, but that the side parts are smaller than those at the top and bottom, and that the dividing is so done that they have flat horizontal surfaces for them to slide on. Behind these side pieces come the wedges, held in place by long bolts which pass through the cap, and these have nuts on them. By screwing down these nuts, the wedges are drawn upwards and thus close the side pieces inward as required. Behind the wedges are metal pads for them to slide on, and in the plan it can be seen that both the pads and the wedges extend right across the steps up to their flanges. As for the rest, it is very similar to an ordinary bearing, only, being a very large one the cap is held on by four bolts, two on either side. This is one of the reasons for making the section line zig-zag. This line, as is seen in the plan, commences on the left-hand side, passing through the center of one of the bolts holding the cap on, and then when half-way across the wedge it suddenly bends downwards, and then bends again into the horizontal, passing through the step up to the vertical center line. Then it again bends downwards until it reaches the horizontal center line along which it next passes right across the plan. It starts then at A, and ends at B. Thus, there are three distinct sections in the elevation. The one furthest to the left, is a section of the part about one of the cap bolts, the next of a part of the wedge and step, and the last of the remainder of the bearing when cut along the horizontal center line as seen in the plan. Each shows just sufficient for the particular plane in which it is, and that is all. Thus a section through any other of the cap bolts would be

Sectional Elevation

Section line

A

B

Sectional Plan

a.

b.

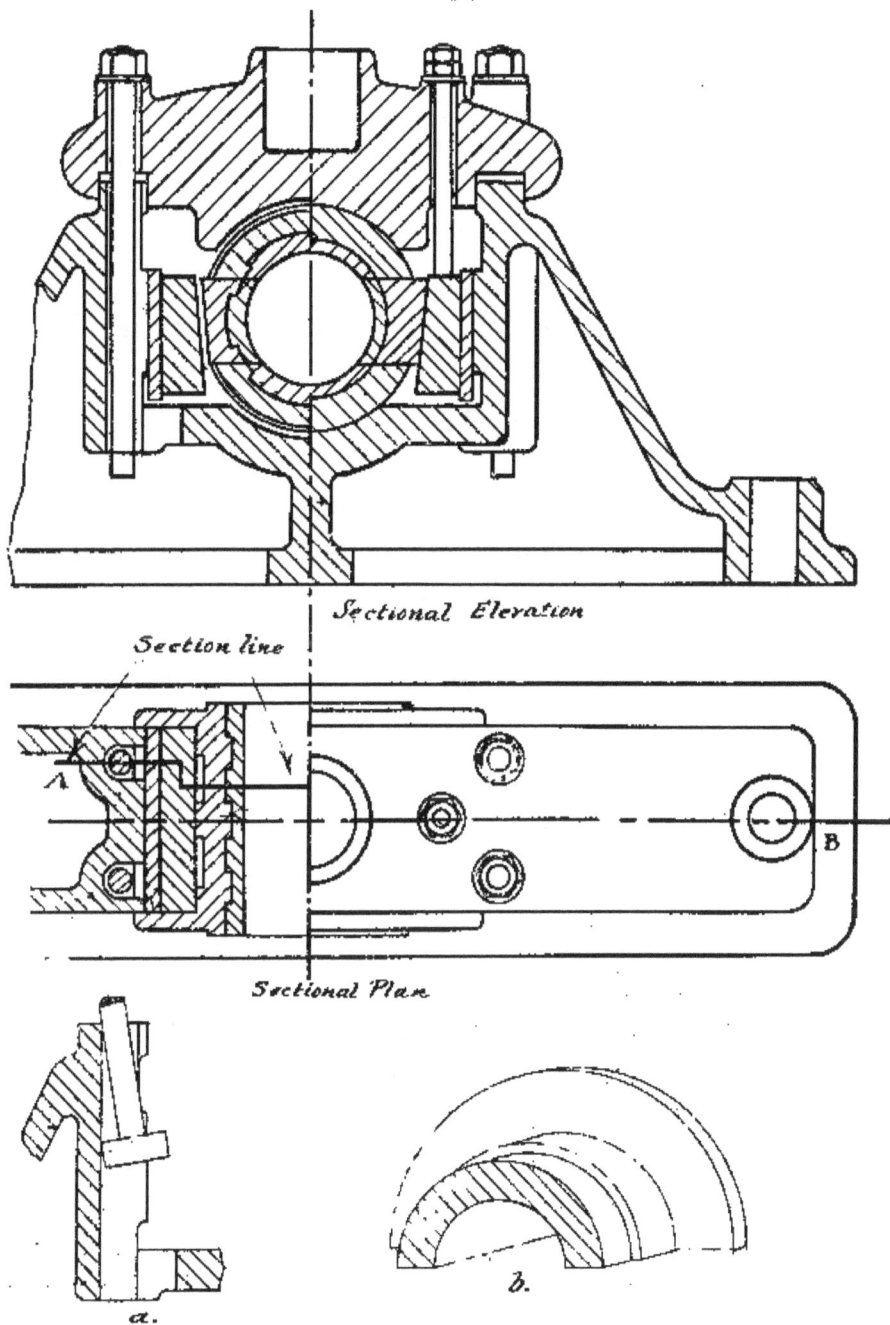

Fig. 12. Sections in more than one plane.

exactly the same as the one shown, and which, moreover, if extended further to the right than it does, would not serve any useful purpose. The same can be said of the other sections, for taking the one on the horizontal center line, then if it be extended to the left of the vertical center line all that is on the right in the elevation would have to be repeated on the left, which would certainly be useless labour.

As for the half-sectional plan, it is a section on the horizontal center line in the elevation, and on that line alone, and so can be readily understood. Returning now to the section through the bolt and using the sectional plan to assist in interpreting it. Firstly, it is seen that the pedestal is hollow. This is shown in the plan by the two horizontal hatched parts to the extreme left. Next it is seen that there are two holes for the cap bolts to pass through, surrounded by solid metal, which the sectional elevation shows to be of the nature of tubes. And these are joined together by a solid piece of metal, as can be seen by the plan. On a close examination in the plan of the holes in these tubes it will be found that they are of a peculiar shape, being semi-circular towards the left and then bounded by two horizontal lines right up to the pad. As a matter of fact, it is open towards the right, that is to say, if the pad and the wedge were removed, there would be nothing to keep the bolts within the tube. The reason for this is that the bolt is passed down the hole from the top, and then, when right through is turned half the way round, so that its head, (which is a tee-head, by the way), catches on the sides of the hole. If the hole were not open on one side, the bolt-head could not pass through it. The bolt on its way down is shown at (a). Fig. 12. That the tube is open on the right side can be seen in the elevation also, for, if it were closed, there would be solid metal to be cut through on this side, and this would be shown as hatched, which is not so shown, and, therefore, the tube must be open. There can be seen, however, on the right-hand side of the tube a narrow, white space, which is the back part of the hole behind the section, and which, because it is seen, shows that a sectional elevation and not a section has been here drawn. The difference between these has been explained on p. 25. It will be noticed that this white space becomes wider behind the pad. This is so because here the part of the pedestal which surrounds the steps, wedges, and pads is thickened up so as to form a bearing strip for the latter. Notice also that in the bottom of this part of the pedestal there is a white patch. This is a hole to allow the bolt head to pass through. It will be seen also that on the left side of the bolt there is a white space between the latter and the hatched part. This shows that the hole is larger than the bolt, that it is, in fact, a clearing hole. And it is confirmed by the sectional plan, for here is a hatched circle showing the section of the bolt, and around this is a white space, showing again that the hole is larger than the bolt. Still looking at the sectional plan, a vertical line will be seen across the hole between the hatched circle and the pad. This is one of the sides of the bolt-head which can be seen at the bottom of the hole. In the sectional elevation notice that, as usual, the bolt is not in section, but in elevation. This has been explained on p. 28. As for the section of the pad, there is nothing peculiar about it. The pad is seen to be a comparatively thin piece of metal, somewhat longer than the wedge, while the sectional plan shows it extending right across the inside of the pedestal. Coming next to the wedge. It must be remembered that the plane of section suddenly alters, and, strictly speaking, there should be in the sectional elevation a vertical line down the wedge, showing where this change of the plane of section takes place. But as this might be misleading it is left out, so the section of the wedge shows simply a piece of metal, which extends across the pedestal as the pad does. This, of course, is seen in the sectional plan. And now, leaving the cap to be dealt with later, the next section is arrived at. Here is shown the remainder of the wedge and the steps up to the vertical center line. Between the wedge and the side step is a white space, which, on referring to the sectional plan, is seen to be a recess in the back of the step, for the section line passes through one of these recesses. Looking next at the parts of the upper and lower steps in this section, three lines are seen to surround them. These show that both the upper and lower steps and the cap and casing containing the steps, etc., are also recessed. Thus the outer arcs (an arc is a part of a circle) are the bottom of the recess in the cap in the one case and in the casing in the other. The next arcs, which are the middle ones of the three sets, are the top of the recesses of the cap, the casing, and the upper and lower steps, and thus show where the cap touches the upper step and the casing the lower one. This is away at the back of the recesses, as can be seen at (b), Fig. 12. And, lastly, the inner arcs are the bottoms of the recesses, in the upper and lower steps respectively. That these recesses exist is confirmed by the fact that the outer arcs end with the cap on the one hand and the casing on the other, while the other two sets of arcs continue right up to the side steps. Of course, all three are in reality on the right-hand side of the vertical center line also; but this section ends at that particular line, and, as will be seen later, the next section does not show the recesses. These recesses are made to minimise the machining required on the steps, etc., for only those parts of the steps which touch the cap on the one hand and the casing on the other require machining. The same can be said for the side steps, which touch the wedges. Thus the cost of a large bearing is materially reduced.

Dealing next with the two inner arcs of the steps, which show the section of the white-metal liner with which the cast-iron steps are lined. On closely examining the sectional elevation, it will be seen that the cast-iron steps are recessed there also, but that in this case the white-metal liner fits into the recesses, and is thus prevented from turning round with the shaft, as it is possible it might do if the friction was great enough. The sectional plan shows that the cast-iron step is recessed crosswise as well, and that, again, the white-metal liner fits into the recesses. In this case the liner is prevented from being forced out sideways. The section has been taken so as to pass through one of the recesses in the cast-iron. This can be seen in the plan. Again, it must be remembered that if this section did not end on the vertical center line, all this should be repeated on the right-hand side of this center line. The third section, commencing, as has been said, at the vertical center line in the elevation, follows the horizontal center line in the plan, and cuts the remainder of the bearing into two parts. Starting, then, with the liner, it will be seen that the inner arc is a continuation of that on the left of the center line, but that the next arc is not so far out as the second one on the left. On looking at the sectional plan the reason of this is seen, for the section being now on the horizontal center line, and this line passing, as it does, through corresponding points on the right and the left sides of the vertical center line, shows that where the section of the liner is now made, the latter is recessed, and hence is thinner in this section than in the preceding one. Thus, by following the horizontal center line in the plan to the left through the sectional plan all points can be found corresponding to those cut by the section line passing along to the right.

In the sectional elevation it will be noticed that there are no recesses in the liner as there were on the left-hand side; which points out that they exist only where the liner is thickest. In this section, also, there is no space shown between the side step and the wedge. Looking at the sectional plan and picking out the corresponding point, namely, the point where the horizontal center line cuts the outside of the step, it will be seen that there the step touches the wedge, and therefore there is no space between them, and so none is seen in the sectional elevation. This elevation shows, also, that there are no recesses in either the cap, casing, or the steps, for there is only one arc instead of three, as on the other side. Also, on careful examination it will be seen that this single arc on the right-hand side is a continuation of the middle one on the left-hand side of the center line; and thus shows that in the former section the upper step touches the cap and the lower one the casing, as far as both of these extend. Before examining the wedge it should be noticed that the side steps have the same slope as the wedges, at those parts where they bear on the latter. Comparing the wedge on the right with that on the left, it will be seen that the former is the same shape as the latter, thus showing that it is of the same section right across. But the upper part of this wedge on the right has a bolt fixed to it, which passes through the cap and there has a nut and check-nut upon it. Looking in the plan it is seen that the section line passes through the bolt, and so it must be shown in the elevation; here again, as usual, the bolt is in outside elevation and not in section, as it should be, strictly speaking. The same can be said of the nuts. Next following the wedge comes the pad, which is exactly the same shape as in the left-hand elevation, and so does not call for any further remarks, except to point out that this shows that the pad also is of uniform section. Next comes the casing, and here the increased thickness behind the pad can more easily be seen. This thickening as has been explained, is to form a bearing for the pad, which can be machined smooth; and by this projecting beyond the rest of the casing it alone need be machined, thereby saving some expense. The same applies to the upper part of the casing where the cap fits on. This section of the casing is in that part of the solid metal which joins up the tubes through which the cap bolts pass (this can be clearly seen in the half-sectional plan), and therefore no bolt is here visible. To the right of this hatched section the outside elevation of one of the tubes can be seen; and that this is perfectly correct is shown by the sectional plan; for here the cap bolts are on either side of the horizontal line and some distance from it. Thus one can be seen in outside elevation when the section is on the center line. At the lower end of this tube can be seen the head of the bolt which passes through it; and as the bolt is a tee-headed one, the head in this view is no wider than the bolt itself. After this the outer casing of the bearing is reached, and this, as is seen, slopes gradually downwards until, with a sudden sweep round, it becomes horizontal. Notice that nothing of this can be seen in the plan. In this horizontal part one of the holding-down bolts is situated. Around this bolt is a short tube cast in one with the casing. The latter is strengthened considerably around the bolt-hole. And this is needful, for often the nuts on these bolts are screwed down very tightly. On the outside of the casing round the bolt-hole a boss is formed, as can be seen in the sectional elevation. This is done that a convenient surface to screw the nut down upon can be obtained. After this the casing suddenly becomes vertical and then as suddenly horizontal again, and here broadens out into a flange on the outside of the casing, which, together with another flange on the inside, forms a wide base for the bearing to rest on. This, of course, extends all the way round the casing. Not, however, that the bearing itself requires this, but the foundation on which it stands, being made either of brick or of concrete, would probably not stand the load which such a bearing puts upon it if the bearing surface were not somewhat extensive. The outer flange can be seen in the plan, being represented by the two outside lines passing round the plan; while the inside flange is seen in the sectional elevations, being shown by the two parallel and horizontal lines across the bottom of the bearing. Nothing has yet been said about the hatched part in the middle of the elevation, which extends down from that part of the casing supporting the steps, etc. This then is a rib extending across the bearing from the outer casing on the one side to that on the other, and which widens out into a flange as shown. This rib adds greatly to the strength of the inner casing, which supports the steps, etc., and thus assists it in carrying the load without distortion. And now there only remains the cap to be examined. The outline of this can be seen in the sectional elevation very readily, while in the plan it is seen to be as wide as the casing. In the elevations it is seen that the top is curved, and that this curve becomes sharp at the ends, thus rounding it off. Next starting from the left-hand side, and remembering that the section is here taken on a horizontal line in the plan through the center of the bolts, first the bolt passing through a clearing hole in the cap is seen; that the hole is a clearing one is shown by the white spaces on either side of the bolt. Here, also, the cap fits on the inner casing—but more of this when the right-hand section is dealt with, for there it can be seen more plainly. There is a boss on the top of the cap for the nut to be screwed down on. This gives a flat surface which could not otherwise be obtained, seeing that the cap is curved here. The boss, of course, is shown in section, for the section has been taken through the middle of it. It must

be remembered that soon after this the section
plane changes, but that here, as in the other part
of the bearing, no line is drawn to show where
this change takes place. As has been pointed out
it serves no useful purpose, and might even be
misleading. The next section shows the recess in
the cap, as has already been explained, and it
ends at the vertical center line. It should be
pointed out that the section of the oil well is not
correctly shown. It is drawn as though it were
symmetrical with the right-hand section; which
it certainly is not, for the plane of section is
behind that for the right-hand portion. If, however, it were drawn correctly, it would look so
very odd, that for preference it is often drawn as
shown. Passing on towards the right—the section
now being on the horizontal line in the plan—the
hole through which the bolt holding the wedge in
place passes is seen. This is again a clearing hole,
as can be seen by the white spaces on each side of
the bolt. On the top is a boss, as there was on
the other bolt-hole, and for exactly the same
reason. After this the cap again fits over the
enlarged part of the inner casing. It will be here
noticed that there is a space between the cap and
the top, of this enlarged portion; again this is
shown by a white space. This space permits the
cap to grip the steps, and thus prevent them from
moving; indeed, this is the function of the cap,

and, as can be readily understood, it could not press the top step down on the others if the bottom of the groove in the cap rested on the top of the enlarged portion of the inner casing. Beside the boss and nuts on the wedge-bolt there is seen another boss and nut; this one, however, is an outside elevation, and thus must be further back than the boss shown in section; otherwise it, too, would be in section. A glance at the plan shows that it is certainly further back than the boss shown in section, for it is the boss for one of the bolts holding the cap in place. Having at last completed the examination of this drawing composed of part sections, it can be seen that, however complicated it may seem at first, yet, when analysed bit by bit, it becomes plain. It is usual to find written beneath such sections such words as these : "A section through AB," or some other line as the case may be. When this is so a search must be made for this line, so that the section can be understood. If, however, a section is shown without any such writing beneath, it may be assumed that it is a section through the center line of the other view. Thus a sectional elevation is a section through the center line of a plan, and a sectional plan is a section through the center line of an elevation. A drawing of this kind saves much valuable time in the drawing office, for two other complete

D

views would be required to convey the information given by these two. Another method of saving time is shown in Fig. 13. Here the front and back elevations of a part of a grip clutch are shown, while below there is a composite view, consisting of half the front, and

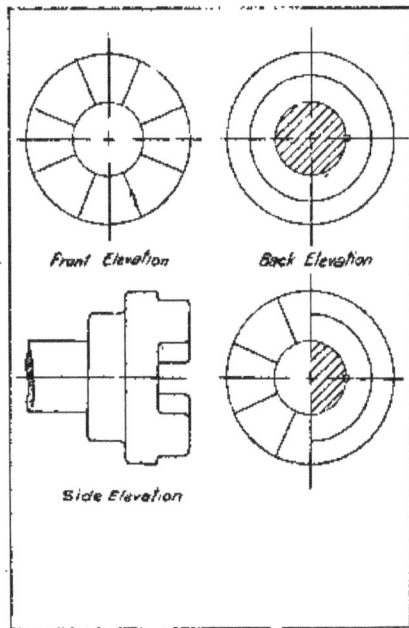

Front Elevation   Back Elevation

Side Elevation

Fig. 13.

half the back elevations. Thus on the left of the vertical center line are shown lines corresponding to those in the front elevation, while on the right of this centre line can be seen the lines corresponding with those in the back elevation. It can be quite understood that such views can only be drawn when the object is symmetrical about the center line, that is to say, that lines appearing on the left of the center line in either front or back elevation appear also on the right in corresponding positions. Sometimes these half views are drawn a slight distance apart, instead of making one view of them. In this case, each has a center line where it would actually be if the view were completed, and not through the middle of the half view. The side elevation or the plan drawn with these composite elevations is the actual view in each case, and not that of the composite view. Thus the plan in Fig 14 would be the same as the side elevation, which is shown, and not be half with the front elevation to the bottom, and half with it to the top, as it would be if the plan of the composite view were drawn. Thus if a view is seen which is unsymmetrical, and from the other views it can be ascertained that this unsymmetrical view is really two views, one from either side of the object, then it must be understood that in reality the object is symmetrical about its center line, and that the view has thus been drawn to save time.

If an object is complicated and also composed of a number of parts, a view is usually drawn showing the object with the various parts in their proper positions, and then sets of other views showing each part separately. The first view is called the general arrangement, while the others are called details. The general arrangement is usually an elevation, either front or sectional, sometimes a plan or side elevation is given with it, but not very frequently. The details are usually shown in both plan and elevation; sometimes a side elevation or a section is added if it will help matters. These latter are fully dimensioned, so that all the sizes can be got from them, while the general arrangement has only a few of the leading dimensions on it, and indeed sometimes none at all. A general arrangement of a connecting-rod end, together with details of parts, is shown in Figs. 14 and 14A. With the latter are also perspective views, so that their shapes can be quickly grasped. Before commencing the examination of these, it is needful that a brief explanation be given of the use of a connecting-rod end, so that the reason for the shape of each part can be understood. A connecting-rod, of which the connecting-rod end is a part, is the link which joins the crosshead pin to the crank-pin. The former of these moves backwards and forwards in a straight line, while the latter describes a circle. And as for the connecting-rod, its motion is very complicated, partaking as it does of both the straight line and circular motions, (see "How a Steam Engine Works," No 5 in this series). To enable the connecting-rod to do this it must have a hole at both ends, the one fitting

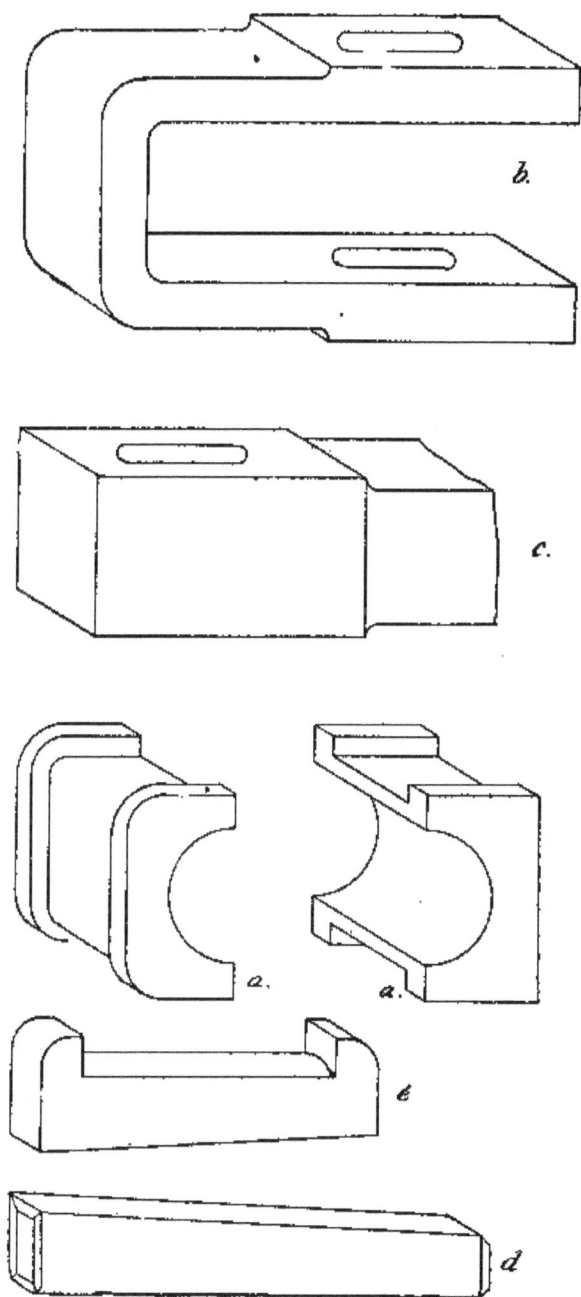

Fig. 14. Perspective of parts of connecting-rod end.

on the crosshead pin, and the other on the crank-pin. Often the situation of one or both of these pins is such that this hole must be made in two pieces of metal, so as to be half in the one piece and half in the other. Then these must be clamped together in place. At the same time it is convenient to take up the wear, that is to say, to make the hole smaller as it gets bigger by wear; being in two parts facilitates this. This, then, is what a connecting-rod end is, a hole at the end of the rod. Returning then to Figs. 14 and 14A, the brasses (a a) will first be examined, they being the principal part, for they contain the hole and all the rest is but a means of fixing them to the connecting-rod. These brasses are drawn a short distance apart to show that they are separate. The hole, then, is of 2 inches radius, that is 4 inches diameter, and in blocks of metal solid up to the dotted lines. Beyond these there are two flanges, one on either side of the brasses, as can be seen in plan, and also in perspective view. They are ½ inch thick, and ¾ inch high. This latter dimension is the difference between 3¾ inches (see elevation), the dimension from the center line to the outside or the flange; and 2⅝ inches (see plan), the dimension from the center line to the bottom of the flange. This is a very common method of giving dimensions, namely all from one fixed line, which is usually a

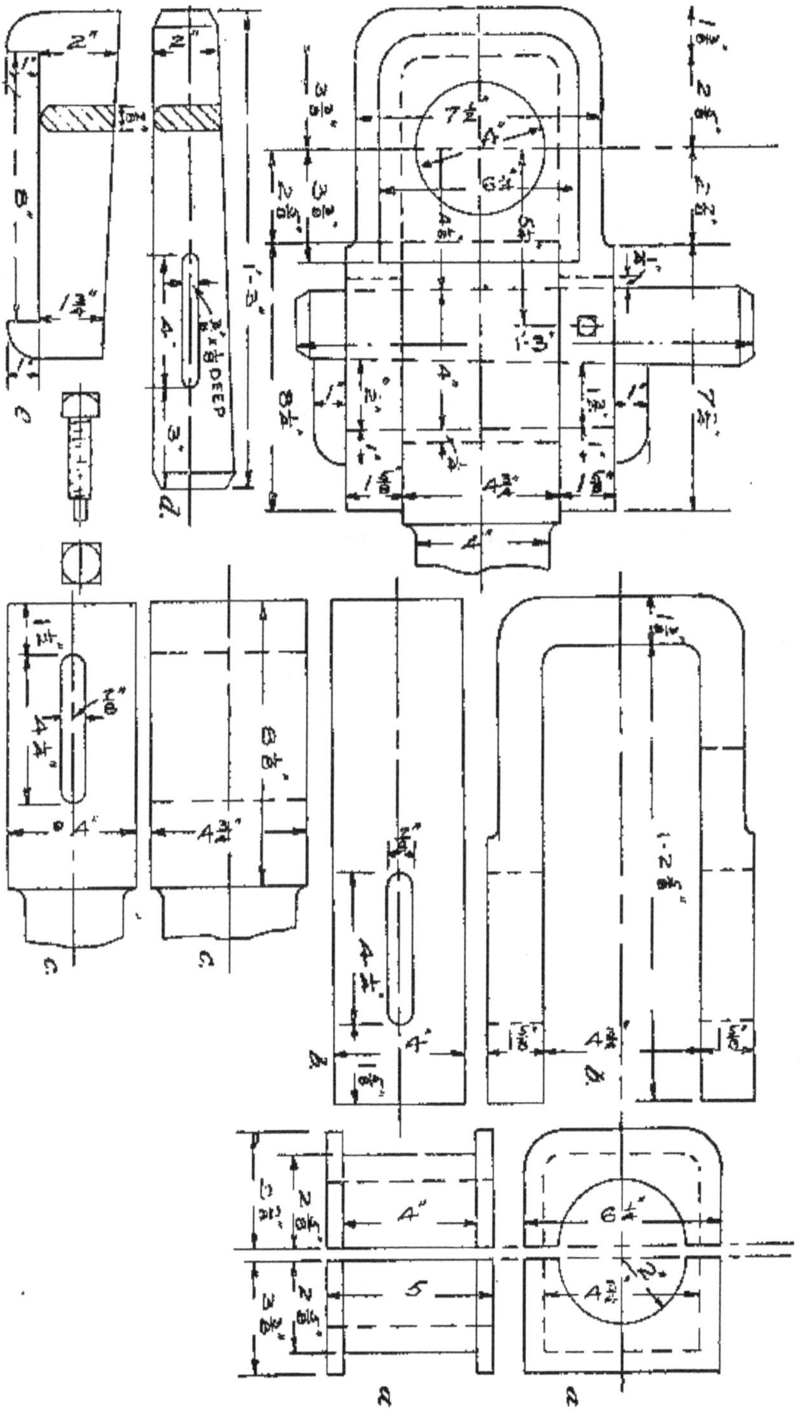

Fig. 14a. A connecting-rod end.

center line. It should be noticed that the dotted lines in the elevation are the bottom of the flanges. These flanges are to prevent the brasses moving sideways. Coming next to the strap, which is a U-shaped piece of metal (*b*) thickened up at the ends, as can be seen in the elevation, and of uniform width, which can be seen in the plan. Each thickened portion is pierced with a rectangular hole, shown as a rectangle in the plan and by dotted lines in the elevation. And indeed it is because of these holes that the legs are thickened up, for otherwise the strap would be weaker here than elsewhere. They are $4\frac{1}{4}$ inches long by $\frac{7}{8}$ inch wide, as is shown in the plan, and pass right through each leg as shown in the elevation.

Dealing now with the other dimensions given for the strap, firstly, it will be seen that the distance between the legs is the same as the distance across the solid portion of the brasses, namely, $4\frac{3}{4}$ inches; also, that the curves at the junctions of the legs with the cross-piece have the same radius as those on one of the brasses (although only one corner is dimensioned, yet it is to be understood that the other one is of the same size, otherwise another dimension would be given). The width of the strap is the same as the distance between the flanges, namely, 4 inches. Because of all these corresponding dimensions, it may be inferred that the strap fits over the brasses, and, moreover, that the latter fit against the cross-piece, which, in fact, is the case, as can be seen in the general arrangement. Now it can be seen how the flanges prevent the brasses from moving sideways, for they embrace the strap. When the brasses are in their proper place in the strap, it will be found that they do not extend up to the end of the latter, for the sum of their lengths is only $5\frac{1}{4}$ inches ($2\frac{5}{8}+2\frac{5}{8}$ inches), while the strap is $13\frac{1}{4}$ inches long. So a search must be made for yet another part to fit into the strap. Examining the other details, it is found that the stub end of the connecting rod (c) is $4\frac{3}{4}$ inches deep, which is the width between the legs of the strap, and also it is 4 inches wide, which is also the width of the strap; so it may be assumed that this also fits into the strap, as the general arrangement shows. Notice, also, that this stub end has a rectangular hole in it of the same size as those in the strap, and that it extends through the stub end, as is shown by the dotted lines in the elevation. So that when the stub end is in its place in the strap there will be a clear way through both. Into this hole it may be guessed the cotter (d) and gib (e) fit, which is practically confirmed when it is found that their thickness is the same as the width of the holes in the strap and stub end. Examining first the gib (e), it will be seen that it consists of a flat piece of metal having two heads, 1 inch high and 8 inches apart. The side between the heads is seen to be parallel with the center line, while the other slopes, so that the gib is thinner at one end than it is at the other. The dimensions show that it is $1\frac{3}{4}$ inches under the head at one end and 2 inches at the other. Thus it tapers $\frac{1}{4}$ inch in 8 inches length, which is the same as 1-32 inch in 1 inch (for divide each by 8, thus—

$\frac{1}{4} \div 8 = \frac{1}{4} \times \frac{1}{8} = 1\text{-}32$ ins., and $8 \div 8 = \frac{1}{8} \times \frac{1}{8} = 1$ in).

This is often written "taper of 1 in 32." There is sometimes a difficulty in understanding this latter expression. It will become clear, however, by examining Fig. 15. Here is shown a taper of 1 in 8, and for convenience we will speak of the units as inches, although all will be equally true for feet, or even millimetres. It will be seen, then, that the upper figure has two lines starting from the same point—one of which is horizontal, the other rises higher and higher at a uniform rate as it recedes from the starting point. At a distance marked 8 inches on the horizontal line, the sloping line is 1 inch above the former. This then is spoken of as a taper of 1 in 8. If the horizontal distance were 16 inches, then the sloping line would have reached a point 2 inches above. For $8 \times 2 = 16$ and $1 \times 2 = 2$, and so on; thus whatever 8 is multiplied by, 1 must be

multiplied by the same quantity. Also the taper
is the same if the lines do not start from one
common point. This is shown in the lower figure.
Here the lines start with a distance of 1 inch
between them, and at the end of the 8 inches
there is 2 inches between them. Thus 1 inch
has been gained, and therefore the taper is 1 in 8
as before. Hence the rule, "Having given the
taper, divide the length of the object by the

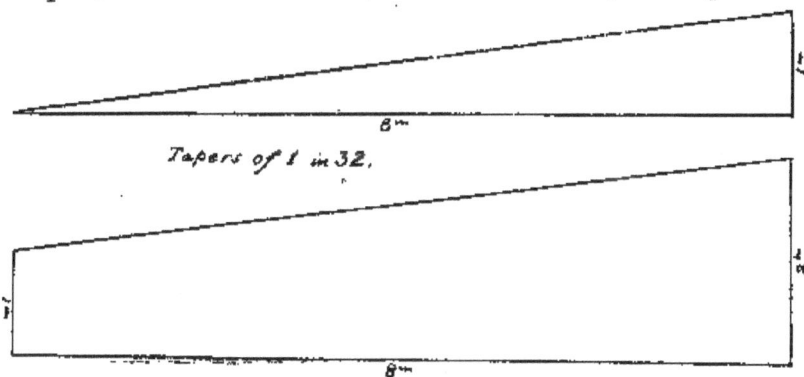

Fig. 15. Tapers of one in eight.

quantity following 'in' given as the taper, and
the quotient is the difference between the dimensions across the ends of this length." Thus,
suppose the length to be 1 foot 1 inch and the
taper 1 in 32; the quantity following "in" is 32;
therefore, divide 13 inches (which is, of course,
the same as 1 foot 1 inch) by 32. This is 13-32*
inches, which, as stated in the rule, is the
difference between the dimensions at the ends of

*13-32 means thirteen thirty-seconds, but for convenience is written in figures,
with either a horizontal or a sloping dash between.

this 13 inches. And thus, if the width at the narrower end is already $\frac{1}{2}$ inch, the width at the broader one is $\frac{1}{2}+13\text{-}32$inch$=29\text{-}32$inches. Sometimes the taper is stated at so much per foot. Take $\frac{1}{8}$ inch per foot as an example. This is the same as 1 in 96, for it is $1\frac{1}{8}$ inch in 12 inches, and multiplying each by such a quantity as will convert the first into unity—in this case by 8—then 1 in 96 (12×8) is obtained as stated.

Returning now to the drawing, and examining the cotter, it will be found that also has one side parallel to the center line and the other tapering. It is stated that the taper is 1 in 32 which is exactly the same as that on the gib. And from this, and the dimension of the narrow end, the width across the broad end can be found. Thus the length is 1 foot 3 inches, that is 15 inches, and dividing this by 32, $\frac{15}{32}$ inches is obtained, and as the width of the narrow end is 2 inches, then that of the other end is $2\frac{15}{32}$ inches.

The only other thing to notice about this cotter is that there is a groove in it 4 inches long $\frac{3}{8}$ inch wide, and $\frac{1}{8}$ inch deep. About this, it should be pointed out that as far as the drawing shows there might be a hole $4\times\frac{3}{8}$ inches right through the cotter, instead of only a groove $\frac{1}{8}$ deep. The last detail to be dealt with is the small bolt shown on the drawing. It is called a set screw and is used to hold the cotter in place.

To do this it is screwed through a tapped hole in the strap, and then pressing on the cotter forces it tight against the back of the hole through which the latter passes. So that the set screw can be withdrawn out of the hole after having been screwed down so tightly, that the end is flattened out, the thread is turned off at the end and its diameter reduced below that at the bottom of the thread. Also when it is screwed against the cotter it is apt to burr the latter, and that is the reason for the groove in the cotter, because then the burr is formed at the bottom of the groove, and thus does not prevent the cotter from being drawn out of its hole.

Coming next to the general arrangement, it has already been found that the strap embraces the brasses and the stub end, also that the gib and cotter pass through both strap and stub end. On this view it should be first noticed that the center lines cross one another at the center of this hole, and that some of the dimensions are given from these center lines. Next it will be seen that much is covered up by the various parts and that dotted lines are used to show this. Thus the flanges on the brasses entirely hide the inner part of the strap, and also a greater part of both the gib and cotter are hidden by the stub end and strap. Notice how the former fit into the latter. The gib fits close against the

end of the holes in the strap, while the cotter fits
close against the gib and against the stub-end.
And at the same time, there is a space between
the ends of the holes in the strap and the cotter,
and also between the gib and the end of the hole
in the stub end. This is very important, for as
the cotter is driven downward it tends to push
the gib over toward the end of the hole in the
stub-end. The gib carries with it the strap, for
it is tight against the latter, and the strap carries
with it the outer brass. This action closes the
brasses together and when this is done it jams
the inner brass against the end of the stub end.
And thereby the hole is made practically solid
with the connecting-rod. The dimensions on
this general arrangement are for the most part
the same as those on the details; and even
when they are not just those required the details
can be obtained from them. Thus, the length of
the hole in the stub-end is not given; but the
width of the cotter and gib taken together is, and
also the space between the gib and the end of the
hole, so that if these be added together the length
of the hole is obtained, namely 4¼ inches. If a
plan be placed beneath this general arrangement,
and the dimensions shown on this for the width
of the various parts, then from these two views
all the details could be made; although for any-
one who had not seen a connecting-rod end the

details would certainly be easier to follow. On the drawing it is seen that the various parts are dimensioned as so many inches, yet it is quite clear that they are not so here. It has been drawn to what is known as another scale than full size. That is to say, a part of an inch has been chosen to represent an inch, and the drawing has been made to this new inch. Thus, if it were made to a scale of 3 inches equal 1 foot, then each inch would be drawn the length which is actually ¼ inch. If 6 inches equals 1 foot then it would be half full size, and ½ inch of a rule would measure 1 inch on the drawing. However, even when the scale is given to which the drawing has been made, it is not safe to use it on either a blue or white print. That this is so will be understood after the following brief description of the making of a blue print. A drawing is first made with a lead pencil on drawing paper. Then over this a piece of tracing paper is stretched, and the drawing carefully inked-in on it by tracing over the drawing beneath. This is called a tracing. This tracing is next put in a frame, called a printing frame, and behind it is placed a piece of paper, on the surface of which are certain chemicals that are, among other things, soluble in water; but after being exposed to the light they become insoluble. Then the frame and contents are exposed to the light. The tracing paper being semitransparent allows the light to pass through it, but where the ink is no light can pass. This light then acting on the chemicals on the surface of the prepared paper changes them, and renders them insoluble; while in those parts of the paper which are protected from the light by the ink on the tracing paper the chemicals remain soluble. After this the prepared paper is taken out of the frame and soaked in water. Then the chemicals are washed out of the paper from those parts where they still remain soluble—that is, where the ink on the tracing kept the light from the paper; while all the rest, having been rendered insoluble by the light, is untouched by the water but turns an intense blue. Thus white lines on a blue ground are obtained. Also, if a white print is made, a similar but more complicated process has to be gone through. Lastly, after having been thoroughly washed, the print is dried. This is sometimes done before a fire. The paper, after being soaked through and then dried, is apt to shrink ; and therefore, however carefully the drawing and tracing may have been drawn to scale, the making of the print spoils all, in as far as the obtaining the correct measurements are concerned. Trust then to the dimensions given, and enquire of someone in authority for those which cannot be found.